EPIC

ANIMAL JOURNEYS

Epic Migrations By Land

SONYA NEWLAND

CRABTREE
PUBLISHING COMPANY
WWW.CRABTREEBOOKS.COM

Author:
Sonya Newland
Series research and development:
Reagan Miller, Janine Deschenes
Editorial director:
Kathy Middleton
Editor:
Ellen Rodger
Proofreader:
Izzi Howell, Melissa Boyce
Graphic design:
White-Thomson Publishing
Katherine Berti
Image research:
Sonya Newland
Print and production coordinator:
Katherine Berti

Images:

Alamy
Minden Pictures p. 9b, 22; Hemis p. 10b; WaterFrame p. 13; Suzanne Long p. 14t; National Geographic Image Collection p. 26; Max Allen p. 29

Getty Images
Ingo Arndt/Nature Picture Library p. 15;

iStock
TSnowImages p. 25b

All other images by Shutterstock

Contents page: Pronghorn are a migratory species that live in North America

Library and Archives Canada Cataloguing in Publication

Title: Epic migrations by land / Sonya Newland.
Names: Newland, Sonya, author.
Description: Series statement: Epic animal journeys | Includes index.
Identifiers: Canadiana (print) 20190098031 |
Canadiana (ebook) 2019009804X |
ISBN 9780778763680 (hardcover) |
ISBN 9780778763727 (softcover) |
ISBN 9781427123497 (HTML)
Subjects: LCSH: Animal migration—Juvenile literature. |
LCSH: Nature—Effect of human beings on—Juvenile literature.
Classification: LCC QL754 .N49 2019 | DDC j591.56/8—dc23

Library of Congress Cataloging-in-Publication Data

Names: Newland, Sonya, author.
Title: Epic migrations by land / Sonya Newland.
Description: New York, New York : Crabtree Publishing Company, [2019] |
Series: Epic animal journeys | Includes index.
Identifiers: LCCN 2019015870 (print) | LCCN 2019017767 (ebook) |
ISBN 9780778763680 (hardcover) |
ISBN 9780778763727 (pbk.)
ISBN 9781427123497 (ebook)
Subjects: LCSH: Animal migration--Juvenile literature. | Nature--Effect of human beings on--Juvenile literature.
Classification: LCC QL754 .N485 2019 (print) | LCC QL754 (ebook) |
DDC 591.56/8--dc23
LC record available at https://lccn.loc.gov/2019015870
LC ebook record available at https://lccn.loc.gov/2019017767

Crabtree Publishing Company

www.crabtreebooks.com 1-800-387-7650

Printed in the U.S.A./082019/CG20190712

Published in Canada
Crabtree Publishing
616 Welland Ave.
St. Catharines, Ontario
L2M 5V6

Published in the United States
Crabtree Publishing
PMB 59051
350 Fifth Avenue, 59th Floor
New York, New York 10118

Published in the United Kingdom
Crabtree Publishing
Maritime House
Basin Road North, Hove
BN41 1WR

Published in Australia
Crabtree Publishing
Unit 3 – 5 Currumbin Court
Capalaba
QLD 4157

Contents

Moving Overland

What Is Migration?

Each year, large groups of animals travel from one place to another and back again. This type of mass movement is known as migration. Many different species of birds and marine animals migrate. But migration is less widespread among land animals. This may be because traveling long distances over land takes a lot more time than moving through air or water. It also requires a lot more energy! Despite this, some land animals make incredible journeys every year.

African elephants usually migrate to regions near rivers or lakes during the dry season.

The Changing Seasons

Most migrations are seasonal. That means animals move to different areas as the seasons change. They might travel to warmer regions to avoid cold winters in their summer territory. Food and water may also be scarce in winter. So animals travel to places where they are more abundant. Every year, elephants in Mali, in northwest Africa, make a long circular migration. They travel 300 miles (483 km) simply to survive, searching for water in the harsh desert environment.

Safe Birthing Grounds

Sometimes animals migrate to the same places every year to mate or have their young. Saigas are a type of **critically endangered** antelope. They live in the **steppe** regions of central Asia. Each spring, female saigas migrate up to 620 miles (1,000 km) in search of good birthing grounds. They find a place with plentiful food and water, and where **predators** are less common. This gives their young the best possible chance of survival.

Saigas are usually found in herds of 30–40 animals. During the spring migration, thousands of females travel together. In summer, they break into smaller groups again.

An Ancient Instinct

Animals have been migrating for thousands of years. But why would they live in habitats that didn't suit them all year round? **Climate change** may be one reason. Long ago, the areas where certain animals lived may have been the right temperature all the time. Over many years, the climate became warmer or cooler in different seasons. Animals started migrating at different times of year to find areas with the right temperature.

If a population of animals grew too large, food would be scarcer. Some animals may have started traveling to other areas for part of the year in order to survive.

Finding the Way

Some species follow exactly the same route to exactly the same place every year. Even animals that have never made the migration before know where to go. No one knows for sure how they do this. Different species have different ways of navigating.

Some use the Sun to work out which direction to go. Others use landmarks such as mountains to keep them on the right track. Still others may use smells to guide their journey.

Earth has a magnetic field. This is the force that makes a compass needle point north. Some animals may have a special sense that uses the magnetic field to keep them moving in the right direction.

American bison migrated across the plains of central North America for thousands of years. Herds of up to 4 million bison once thundered across the landscape. Overhunting in the 1800s reduced their numbers. Today, only around 15,000 roam freely.

Danger On the Way

During these long journeys, animals have to overcome many obstacles. Often, these problems are caused by humans. Roads built across open areas break up **migration corridors**. In southern California, mule deer and mountain lions used to migrate seasonally. Now, highways, houses, and golf courses have split up the land. This has forced the animals to change their migration patterns.

Climate change caused by humans is also affecting animal migration. As our planet warms up, the areas where different animals can live are changing.

Some desert ants, such as the Saharan silver ant, use the Sun to help them find their way.

Prairie rattlesnakes in western Canada migrate through agricultural land, where they are at risk of being killed by farm machinery.

Porcupine Caribou

Habitat Arctic **tundra** in Alaska, U.S., and Yukon, Canada

Known for Longest migration by a land mammal

When Spring through fall

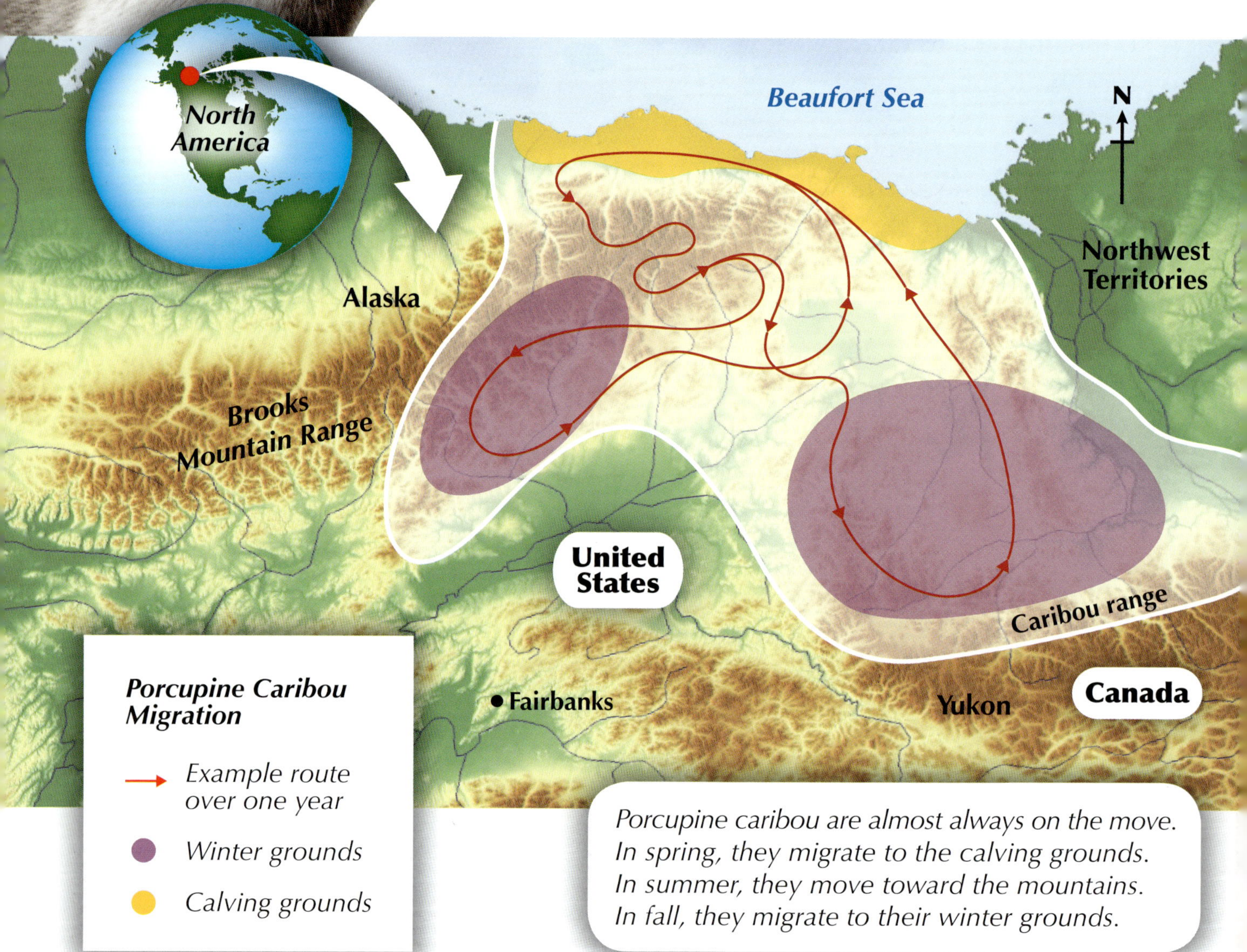

Porcupine caribou are almost always on the move. In spring, they migrate to the calving grounds. In summer, they move toward the mountains. In fall, they migrate to their winter grounds.

The Journey

Each spring, about 218,000 Porcupine caribou set off on an incredible journey through the freezing arctic tundra. Their migration can be more than 3,000 miles (4,830 km) long—the longest of any land mammal. But the trek is worth it. The coastal plains by the Beaufort Sea are the perfect place to have babies, or calves. This habitat has plenty of plants for the caribou to eat. The flat land also helps the herd watch out for predators that may eat their calves, such as wolves and grizzly bears.

Female caribou have their calves in late May and early June. By the time summer begins, the herd has started to move away from the coast to escape the hordes of pesky mosquitoes. Then, in fall, they begin the long journey hundreds of miles south to their winter grounds. The caribou's favorite food is **lichen**, and they eat as much of this as possible before winter.

Only about half of the calves born will survive until the next year.

*Wildlife and government organizations put **GPS tracking** collars on some caribou so they can track the herd's migration.*

Changing Paths

Porcupine caribou are constantly moving, looking for food. This keeps the habitat healthy because they do not eat all the food in one place at one time. But GPS tracking shows the caribou's route changes from year to year. This is often because of changes in their habitat.

Global warming causes some plants to grow for longer periods, and in new places. Caribou move to areas where more plants grow. However, shrubs, trees, and other plants may block the usually clear paths of migrating caribou. Fires can also clear out plants and help new plants to grow. All these factors may change where caribou travel for food.

Escaping Threats

Porcupine caribou face predators, difficult landscapes, and freezing environments. But one of their most annoying threats is mosquitoes! These bloodsucking bugs can bother the caribou so much that they are too distracted to eat, or become unable to nurse their calves. The caribou move away from the coast in summer to escape from them. They travel with their new calves through the mountains, where it is too windy for mosquitoes. They are also less likely to get bitten in large groups.

Reindeer lichen is a common food source for caribou.

The Porcupine caribou population could become endangered if the place where they give birth is damaged or closed off.

Human Roadblocks

Porcupine caribou are facing a new threat. A lot of the caribou's calving habitat is inside the Arctic National Wildlife Refuge (ANWR). In 2017, the U.S. government passed a **bill** that opened up some of ANWR to oil and gas drilling. Drilling could force the caribou out of their safe calving habitat, putting the population at risk.

Habitat Protection

In local **Indigenous** communities, such as the Gwich'in and the Tr'ondëk Hwëch'in, the caribou is an important food source. Many people in these communities want to protect the caribou habitat. Some fight against oil and gas exploration. Environmental groups also work to protect the caribou and their habitat.

The Trans-Alaska pipeline crosses part of the caribou's range. It carries oil from Prudhoe Bay, by the Beaufort Sea, 800 miles (1,300 km) to Valdez in southern Alaska.

Red Crab

Habitat	Mainly rain forest, Christmas Island, Australia
Known for	Yearly migration from forest to ocean in mass numbers—more than 50 million
When	Start of wet season, between October and December

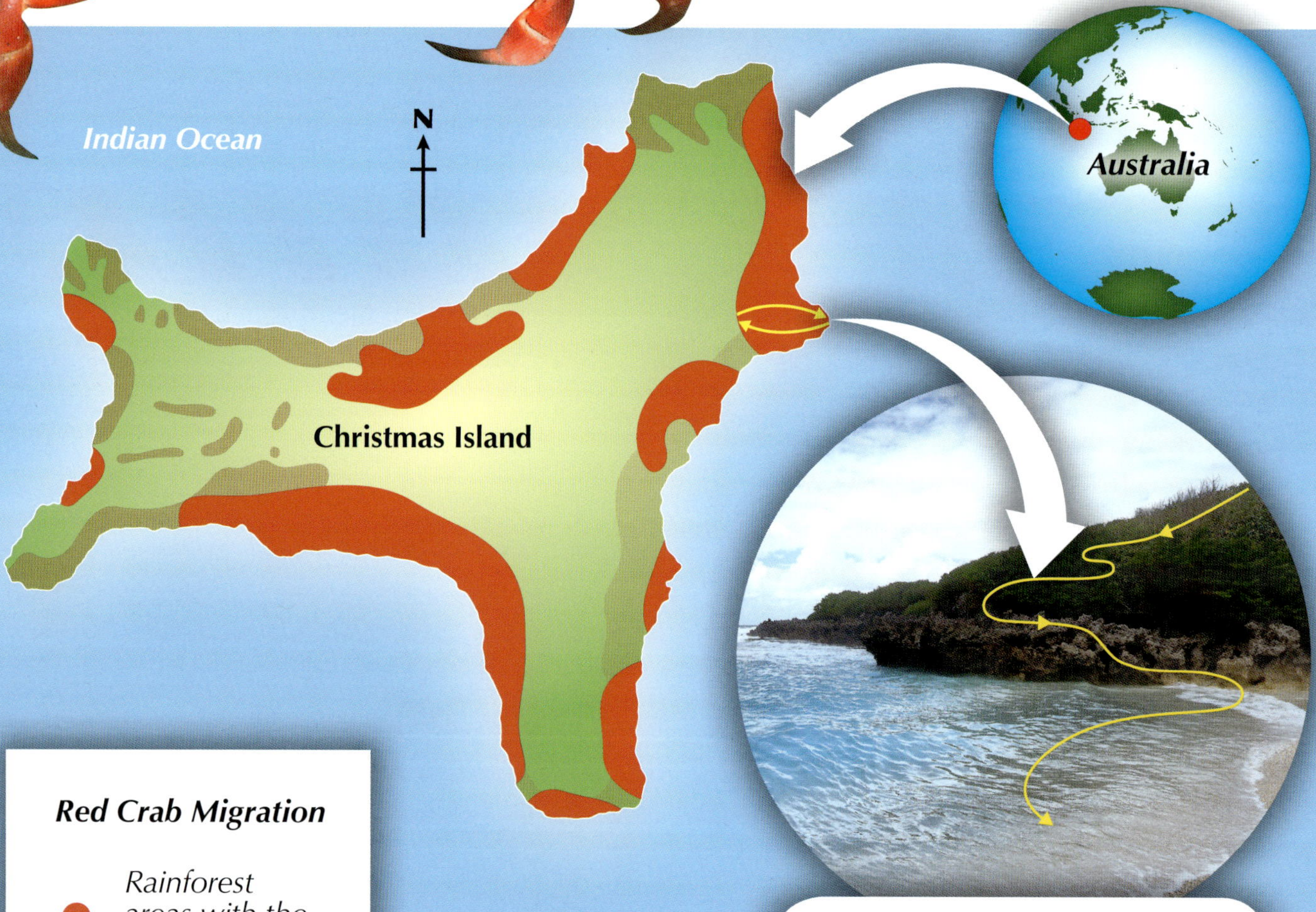

Red Crab Migration

Rainforest areas with the most red crabs

Example route of a red crab

Red crabs leave their rainforest habitats and climb down rocky cliffs to reach the shores of the Indian Ocean.

A female crab may lay as many as 100,000 eggs in the ocean. For a few days, the beaches are covered in tiny crabs.

The Journey

Millions of crabs migrate together on a journey that lasts between 9 and 18 days and covers an average of 5 miles (8 km). At the start of each wet season, the crabs dig their way out from their **burrows** on the rainforest floor. The male crabs leave first. They climb down steep, rocky cliffs to the shores of the Indian Ocean, where they dig burrows in the sand. When females arrive a few days later, they mate with males in the burrows. Then, males begin migrating back to the rain forest. The females stay behind. They lay their eggs in the shallow waters on the coast before returning to the rain forest.

After about a month, the baby crabs have grown big enough to leave the water. They make their own journey to the rain forest. Along the way they have to battle obstacles such as cars and predators.

Don't Dry Out

Red crabs must stay moist to survive. Their burrows are below ground, away from the Sun. They migrate during the wet season to take advantage of the rain, which helps keep them moist. As they migrate, the crabs move quickly to limit their time in the sunlight. When they reach the ocean, they dip themselves in the water to regain moisture.

Migrating crabs take on all obstacles. They climb up and down rocky cliffs and over debris on the forest floor. They cross busy roads—and sometimes wander through people's homes!

Watch for Ants

The biggest threat to red crabs is the yellow crazy ant. Red crabs are prey for this **invasive species**. The ants spray the crabs with a type of acid, which blinds them and kills them within a few hours. Yellow crazy ants can also take over the crabs' burrows. Without their cool, moist homes, the crabs dry out and die.

As they make their journey from the beach to the rain forest, baby crabs are especially vulnerable to yellow crazy ants.

Human Roadblocks

Red crabs cross many obstacles on their journey to the coast. One of their biggest obstacles is vehicles on roads. Many crabs have been killed by traffic that cuts through their migration path. Their rainforest habitat has also been threatened by forest clearing. People have cut down trees to make room for mining. Luckily, this is no longer allowed.

Helping Hands

Assisting the red crab migration is a community effort on Christmas Island. Workers stand on the roads and use rakes to guide crabs safely across. People have also built fences, tunnels, and bridges to help the crabs avoid roads.

Some roads are closed during the migration to let the crabs cross safely.

Blue Wildebeest

Habitat	Grassy plains of southern and eastern Africa
Known for	The "Great Migration"—a circular migration through Tanzania
When	Year round

Blue wildebeests follow a route around the Serengeti National Park in Tanzania. At the northernmost point, they cross into Kenya.

There are so many animals on the migration that herds may stretch for 25 miles (40 km) across the plains.

The Journey

Each year, the blue wildebeests of eastern Africa make an incredible 1,000-mile (1,600 km) migration. They are following the rains around the dry grasslands on a journey of survival.

The migration begins in the south of the Serengeti National Park in Tanzania. Between January and March, half a million wildebeest calves are born here. As summer approaches, food and water grow scarce. Herds of wildebeests begin to move north.

Around August, the wildebeests cross into Kenya. They remain there until the rains begin again in October. Then, they set off southward to their calving grounds once more.

Experts believe that the wildebeests can smell rain. By following this scent, they find areas with more water to drink and plants to eat. Rain clouds and the sound of thunder in the distance can change the path of the herd's migration.

Wildebeests do not follow a leader. This means that smaller herds often break away from the huge main herd. They go in different directions, circling the main group. By the time the herds return to their starting point, the migration may have covered nearly half the Serengeti region!

To help protect young wildebeests from predators such as lions and hyenas, adults keep the calves in the middle of the herd.

Zebras tear up the long grass with their long front teeth, "mowing" it for the wildebeests, which prefer to graze on short grass.

A Huge Group

Around 1–2 million wildebeests take part in this "Great Migration." But they don't travel alone. About 200,000 zebras go too. Wildebeests and zebras are good traveling companions because they do not compete for food. The two species also help each other in different ways. Zebras have better memories than wildebeests. They remember which parts of the journey are safe or dangerous. But wildebeests are better at finding water, which helps the zebras.

Working Together

Lions, hyenas, cheetahs, and crocodiles all lie in wait for the wildebeests. To outsmart them, wildebeests use "swarm intelligence." This is when animals work together to overcome obstacles. The wildebeests move in a curve formation with a wave at the front, like a swarm of bees. Even so, about 250,000 wildebeests and 30,000 zebras die during the migration each year. Some are caught by predators. Others die of hunger, thirst, and exhaustion.

At the Grumeti River crossing, the wildebeests gather on the riverbank. They wait until it seems safe, then cross in large groups. Still, many wildebeests are caught by crocodiles.

Thousands of tourists come to watch the herds. The roads, fences, and lodges built to cater to them often block the traditional migration route. This changes the pattern of the migration.

Human Roadblocks

Predators are part of life for wild animals. But humans are also causing problems for wildebeests. Human settlements in and around the Serengeti force the herds to find different routes. Fences stop the migrating herds from reaching their wet-season lands altogether. When this happens, wildebeests may die from lack of food or water.

Keeping the Way Clear

In some areas that the migration passes through, communities and wildlife organizations are working together to help the animals. They raise money to help pay landowners to keep their land clear of fences and buildings. They also teach people to manage the land **sustainably**. For example, they are encouraged not to **overgraze** their livestock. This will make sure there is enough food for the huge herds.

Emperor Penguin

Habitat	Antarctic ice
Known for	"Tag-team" migration several times a year to and from nesting grounds
When	March to December

The distance between the sea and the nesting grounds can be up to 100 miles (160 km). The male and female penguins may make this trek three times each in a single year.

The Journey

The "march of the penguins" begins as the sea ice breaks up at the end of summer in Antarctica. Emperor penguins live in **colonies** scattered around the edge of Antarctica. They migrate up to 100 miles (160 km) to and from their nesting sites, or rookeries.

Toward the end of March, male and female emperors journey together inland to their rookery. When they arrive, the females lay a single egg. They pass this to the male, who tucks it in a special pouch between his legs to keep it warm and safe. In June, the females begin the long journey back to the sea to find food. They feast on fish, **krill**, and squid before returning.

In August, while the females are away, all the eggs hatch. The male penguins care for the young. They feed the chicks with a milky substance they **regurgitate** from their own bodies. The females return just as the males are running out of food for the chicks. The timing is perfect. They take over caring for their babies while the males make the exhausting journey back to the sea to feed.

Throughout the long, dark Antarctic winter, emperor penguins repeat this journey over and over again. In the final migration of the annual cycle, the whole family returns to the sea. It's a life-or-death journey for the youngsters. Many do not survive.

Experts think this migration and life cycle has evolved to ensure that the chicks reach the sea for the first time at exactly the right time, when food is most abundant.

The males have to move carefully in rotation, balancing the egg on their feet.

Loving Parents

Winter temperatures can drop to –76°F (–60°C) and winds can reach 100 miles per hour (160 kph). To stay alive, the male penguins work on a rotation system. The penguins on the outside of the huddle gradually move inward as those in the warmer middle move outward. In all this time, they have nothing to eat. By the time the females return, the males may have lost up to 40 percent of their body weight.

The Changing Landscape

All across the Antarctic, the ice melts and shifts. This means that at each stage of migration, the landscape has changed. The penguins must find different routes to the sea every time, to avoid obstacles that have appeared. Despite this, they never get lost. And even among the thousands of penguins and chicks in the rookery, a returning penguin will identify his or her own family by their calls.

As the planet warms up, the Antarctic sea ice is becoming thinner. Cracks appear more often, which can change the course of the penguin migration.

Human Roadblocks

Emperor penguins face predators in the sea, such as leopard seals. There is also danger in the air. Some birds will swoop down and take chicks shortly after birth. But human activity is also a threat, even in the remote Antarctic. Climate change is the biggest issue. Melting ice reduces the penguins' habitat. One of the emperor penguins' main sources of food is krill. Young krill eat algae that live in the sea ice, so less ice means fewer krill.

Pronghorn

Habitat	Open grasslands of North America
Known for	Extremely difficult migration path
When	November to April

For more than 6,000 years, pronghorn have migrated from Grand Teton National Park in northwestern Wyoming to the Green River Basin in the south of the state.

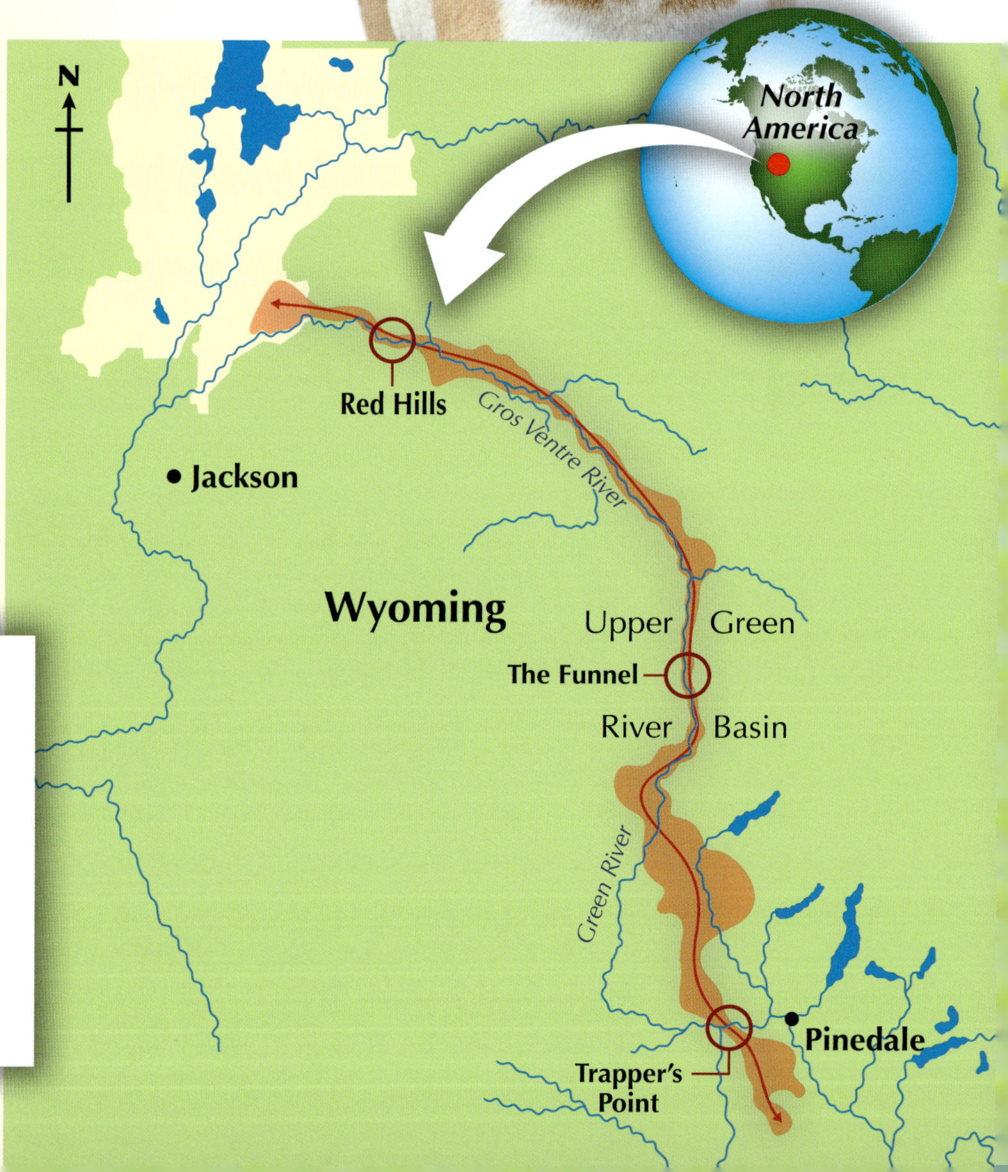

Pronghorn Migration

Grand Teton National Park

Migration corridor

Migration route

Bottleneck

The Journey

Herds of pronghorn can be found in several areas in North America. But one group is famous for making a very difficult migration every year. This herd of 300–400 pronghorn spends its summers in Grand Teton National Park, in northwestern Wyoming. The pronghorn's favorite food, sagebrush, is plentiful there. In winter, temperatures drop and deep snow buries the sagebrush and grass.

Every fall, the pronghorn set off on a 150-mile (241 km) trek southward to warmer wintering grounds. This journey takes them across private land. They have to overcome miles of fencing and travel close to dangerous roads. The pronghorn must also cross four major rivers and a high mountain pass!

The fall migration is a race against time. In the winter, snow blocks several parts of the route. If the pronghorn can't get across before the snow builds up, they may have to spend the winter too far north. When they finally reach the Green River Basin, this herd is joined by others from all over Wyoming. They spend the winter there together. In April, they make their way northward again as the snow melts.

The sagebrush of the Grand Teton National Park provides the pronghorn with food throughout the summer.

Female pronghorn usually give birth to twin fawns.

The "Path of the Pronghorn"

Pronghorn are the fastest land mammal in North America. They also have sharp eyes to see danger in the distance. Their speed and good eyesight keep them safe from predators. Their migration path avoids woodland and swamps. Instead, they migrate across the high land between the woodland and the rivers.

They follow the same route year after year. Even if they come across obstacles, they won't change their path. If the route ends up being blocked completely, they may stop migrating altogether.

Bottlenecks

There are several bottlenecks on the route where human settlement or natural features have made the path hard to navigate. Some areas are as narrow as 330 feet (100 m). That's not much space for 400 pronghorn to get through!

Bottlenecks make the pronghorn vulnerable to predators such as coyotes and wolves. If they don't make it through before snow blocks their path on the fall migration, they may die on the mountain.

Pronghorn can't jump very high, so they have to scramble under fences. Some get caught in the barbed wire and die.

A wildlife bridge has been built for the migrating pronghorn at a bottleneck called Trapper's Point. It gives the animals a clear, safe path across a highway.

Human Roadblocks

Roads, housing developments, and fences to mark private land have been built in the migration area. The pronghorn have to make their way around buildings and across highways.

Fixing the Problem

The Wildlife Conservation Society is one group working to preserve the pronghorn's migration corridor. It monitors the migration to watch for changes. It also works with local authorities to conserve important areas. The Path of the Pronghorn Conservation Program has helped too. The program has removed or adapted more than 120 miles (193 km) of fencing along the route.

Conclusion

An Important Journey

Migrations are key to the survival of many species. But climate change, buildings, and land privatization are all affecting migrations. If animals cannot move freely between summer and winter ranges, they may die of cold or starvation. If they cannot reach safe birthing grounds, young animals may not survive.

Conservation Areas

Some migration corridors are now protected as conservation areas. This means that individuals cannot build houses or farms on them. In Africa, a migration even bigger than the wildebeest migration takes place each year. Thousands of plains zebras travel more than 300 miles (483 km) between Botswana and Namibia. The Kavango Zambezi Transfrontier Conservation Area was set up to protect the zebras, as well as other migrating animals. Now the zebras do not have to get past fences or other obstacles.

Plains zebras migrate in a protected area between Namibia and Botswana. This protected corridor is in the largest conservation area in the world.

Animals such as mountain lions may be fitted with GPS collars, which allow wildlife experts to track their movements using satellites.

Keeping Track

By tracking animals, experts can see how migration paths change each year. This helps them to identify problems.

Mule deer in the U.S. migrate up and down the mountains. Tracking showed that they were searching for certain plants. Tagging also revealed that mountain lions followed the same migration path.

The mule deer were following their food source. And so were the mountain lions! Changes to the vegetation, because of buildings or climate change, were affecting the mule deer and the mountain lions, which hunted the deer for food.

What Can You Do?

Your own actions could be affecting animal migrations. Human behavior is causing global warming, which is one of the biggest threats to migrations. You can help by taking steps to stop global warming. Turn off lights, reduce the amount of water you use, and reuse and recycle instead of throwing things away. This will help slow the changes to animals' environments.

Glossary

bill A proposal for a new law that will be discussed by a government

burrow A hole or tunnel underground that an animal lives in

climate change The change in normal weather patterns in an area over a long period of time

colony A large group of a particular species of bird or animal that lives together

critically endangered Close to dying out completely and becoming extinct

global warming The rising average temperature on Earth

GPS tracking Using satellite global positioning system (GPS) technology to follow animals on the move

Indigenous Describing people who are born in a particular area

invasive species Types of plants or animals that are not native to a particular area

krill Tiny sea creatures similar to shrimp

lichen A living, plantlike substance that often grows on the surface of rocks and trees

migration corridor The route that animals take when traveling between winter and summer habitats

overgraze When livestock eat so much grass and other plants in an area that it damages the soil and the plants do not grow back

predator An animal that hunts other animals for food

regurgitate To bring food back up into the mouth after it has been swallowed

steppe Large, treeless grasslands in eastern Europe and central Asia

sustainably A way of doing things that makes sure natural resources are not depleted

tundra A flat, treeless area in the Arctic

Find Out More

Books

Brundle, Harriet. *Land: Amazing Migrations*. Booklife, 2018.

Franchino, Vicky. *Animal Migrations*. Scholastic, 2015.

Kalman, Bobbie. *Why Do Animals Migrate?* Crabtree Publishing, 2009.

Websites

https://bit.ly/2XzlA44
Find out more about Africa's greatest migration on the National Geographic's page about the wildebeest migration.

https://bit.ly/2X2VBxU
Discover some of the world's greatest migrations by land, sea, and air, including wildebeests and red crabs.

www.youtube.com/watch?v=L7tWNwhSocE
This is the trailer for the 2005 film *March of the Penguins*, about the amazing emperor penguin migration.

Index